PLANETS IN OUR SOLAR SYSTEM

Learn About URANUS!

BY DANIELLE HAYNES

Enslow PUBLISHING

DISCOVER!

Please visit our website, www.enslow.com. For a free color catalog of all our high-quality books, call toll free 1-800-398-2504 or fax 1-877-980-4454.

Cataloging-in-Publication Data

Names: Haynes, Danielle.
Title: Learn about Uranus! / Danielle Haynes.
Description: New York : Enslow Publishing, 2024. | Series: Planets in our solar system | Includes index.
Identifiers: ISBN 9781978533677 (pbk.) | ISBN 9781978533684 (library bound) | ISBN 9781978533691 (ebook)
Subjects: LCSH: Uranus (Planet)–Juvenile literature.
Classification: LCC QB681.H394 2024 | DDC 523.47–dc23

Portions of this work were originally authored by Greg Roza and published as *Uranus: The Ice Planet.* All new material in this edition authored by Danielle Haynes.

Published in 2024 by
Enslow Publishing
2544 Clinton Street
Buffalo, NY 14224

Designer: Andrea Davison-Bartolotta
Editor: Danielle Haynes

Photo credits: Cover (Uranus), p. 1 (Uranus) NASA images/Shutterstock.com; cover (top), p. 1 (top), series art (backgrounds) Jurik Peter/Shutterstock.com; series art (lens flare) andruxevich/Shutterstock.com; p. 4 Natee Jitthammachai/Shutterstock.com; p. 5 PK Designs/Shutterstock.com; pp. 7, 11 buradaki/Shutterstock.com; p. 9 shooarts/Shutterstock.com; pp. 13, 21 NASA/JPL; p. 15 mr.Timmi/Shutterstock.com; p. 17 Mevan/Shutterstock.com; p. 19 (main) Siberian Art/Shutterstock.com; p. 19 (inset) NASA/JPL/USGS; p. 20 courtesy of NASA.

Printed in the United States of America

CPSIA compliance information: Batch #CS24ENS: For further information contact Enslow Publishing, at 1-800-398-2504.

CONTENTS

Discovering an ICE GIANT

The seventh planet in our solar system is called Uranus. Because it's so far from the sun, it doesn't get very much heat. It's known as an "ice giant" because it's so cold and icy. Scientist William Herschel discovered Uranus in 1781.

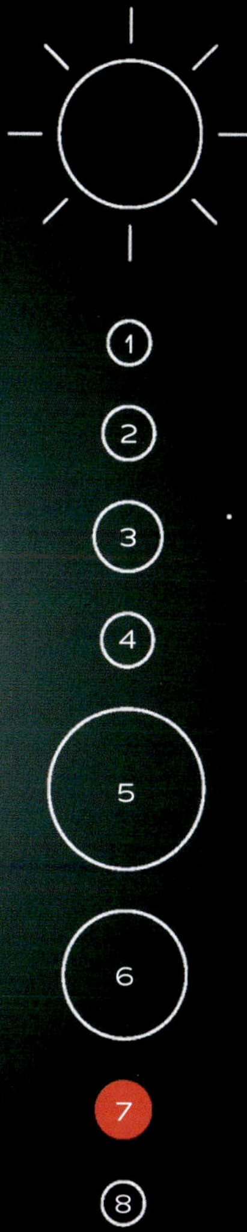

Uranus is usually about 1.8 billion miles (2.9 billion km) from the sun.

See the RINGS

Uranus is the third-largest planet in the solar system. It's about four times as wide as Earth.

Uranus has rings that are very hard to see. There are two sets of rings—an inner set of nine and an outer set of two rings.

Some of Uranus's rings are **surrounded** by dust.

Getting a MOVE ON

Uranus **orbits** the sun, just like the other planets do. Since it's so far away, Uranus takes a little more than 84 Earth years to orbit the sun! Only one planet takes longer. Neptune orbits the sun about every 165 Earth years!

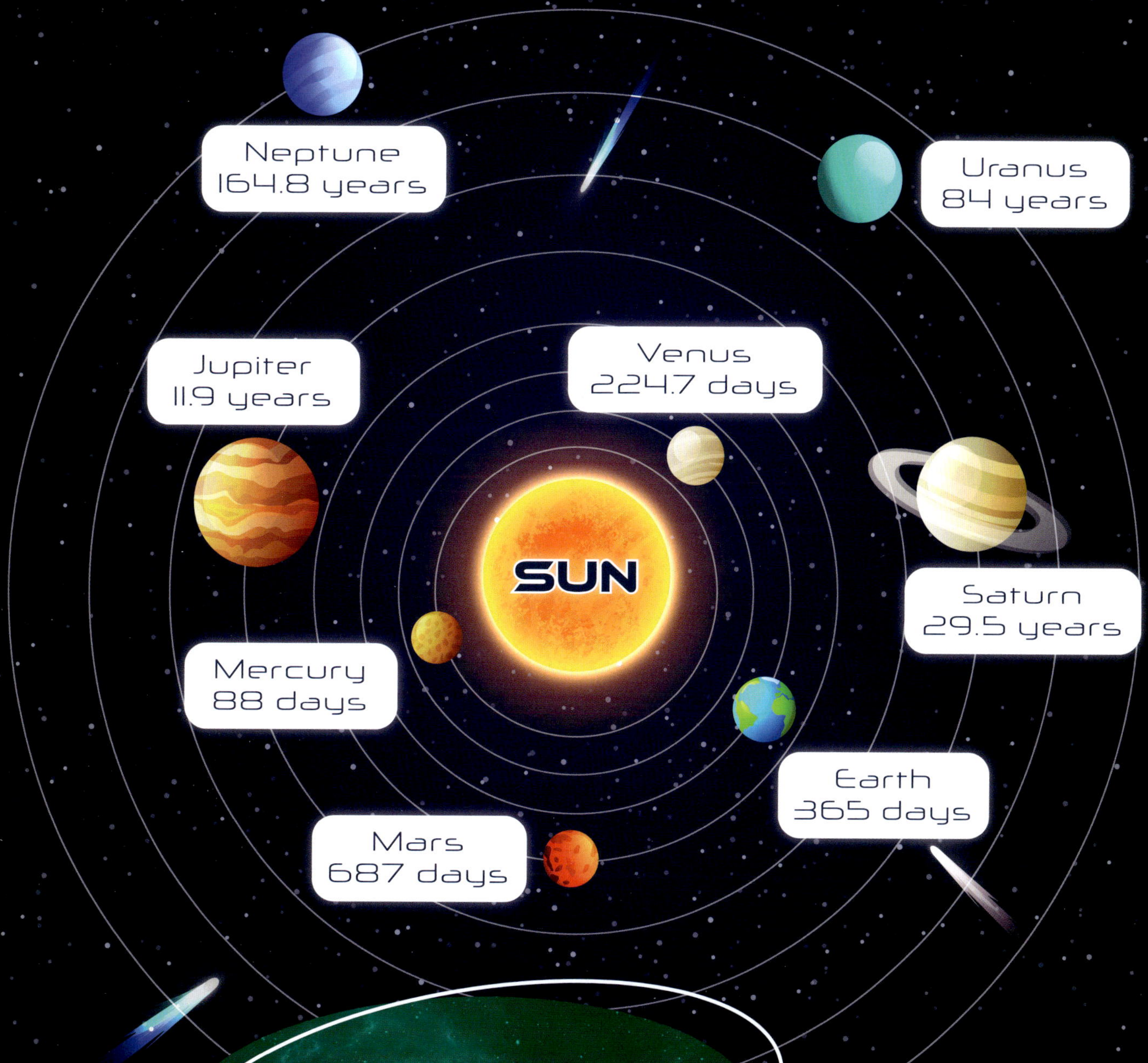

The length of each planet's orbit is shown here.

Uranus spins on its side! Scientists think another planet may have crashed into Uranus long ago. The crash might have tipped Uranus on its side. Uranus makes a complete spin about once every 17 Earth hours. The air around Uranus spins faster at its poles than at the **equator**.

Uranus and Venus spin in the **opposite** direction of other planets.

Long SEASONS

Uranus has seasons, but its tilt makes these seasons very different than on Earth. For about one-fourth of each year on Uranus, the sun shines over each pole. This means that half the planet experiences a winter in darkness that lasts 21 Earth years!

Uranus's orbit is almost a perfect circle, so it's about the same distance from the sun during its entire year.

Icy BLUE

Uranus is the coldest planet in the solar system. Its average temperature is –353°F (–214°C). That's why it's known as the Ice Planet. Uranus looks completely blue. This is because of a gas around Uranus called **methane**.

The gas methane in Uranus's **atmosphere** makes it look blue.

Hidden FROM VIEW

Scientists think Uranus might have a huge ocean beneath its icy clouds. They also think the **core** of the planet is small and rocky. The center may also be much hotter than the planet's icy **surface**.

Uranus's surface is mostly fluid, or like water.

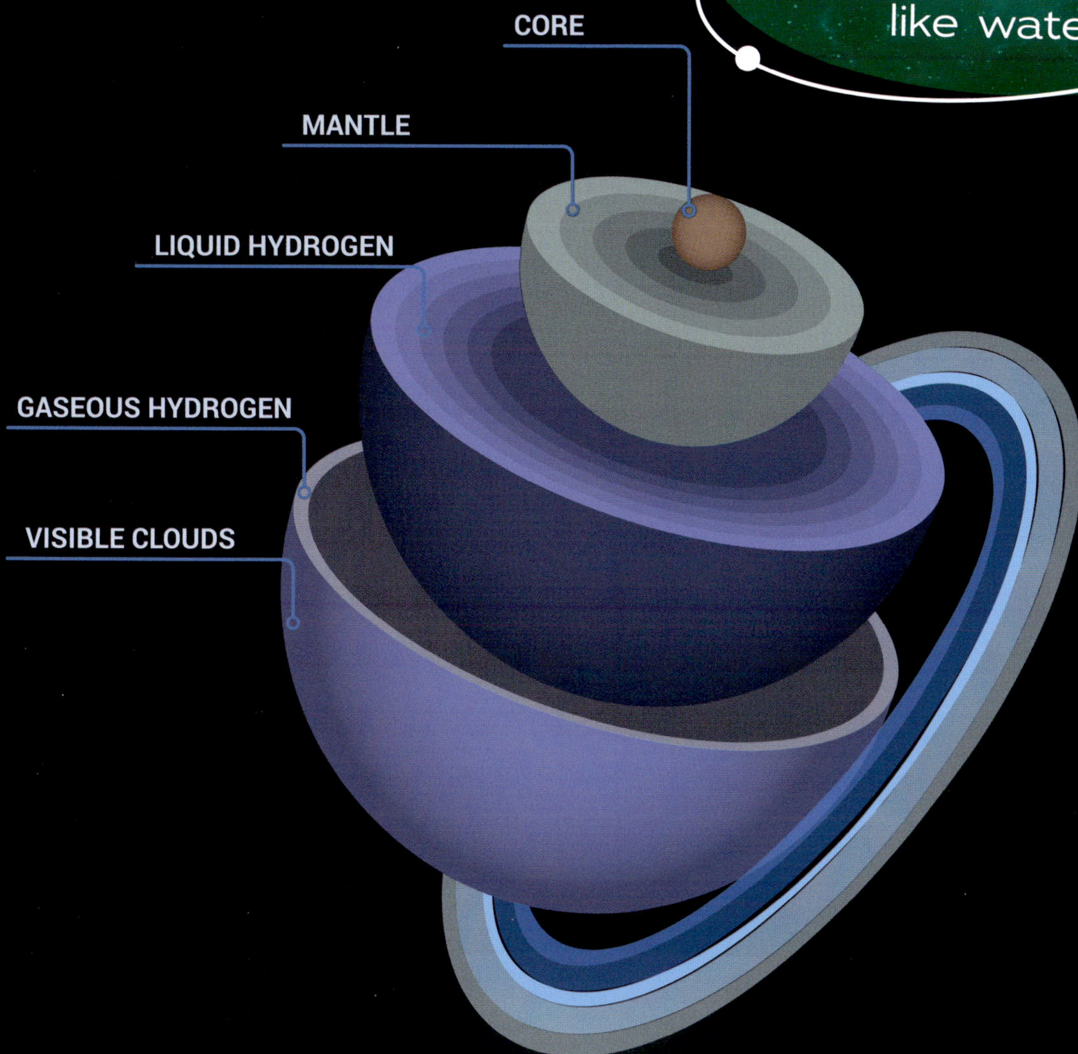

Moons AND RINGS

Scientists have used **telescopes** and **probes** to learn about the Ice Planet. They have learned that Uranus has at least 27 moons. The moons orbiting closest to Uranus are made of ice and rock. Some of Uranus's moons might even have underground oceans!

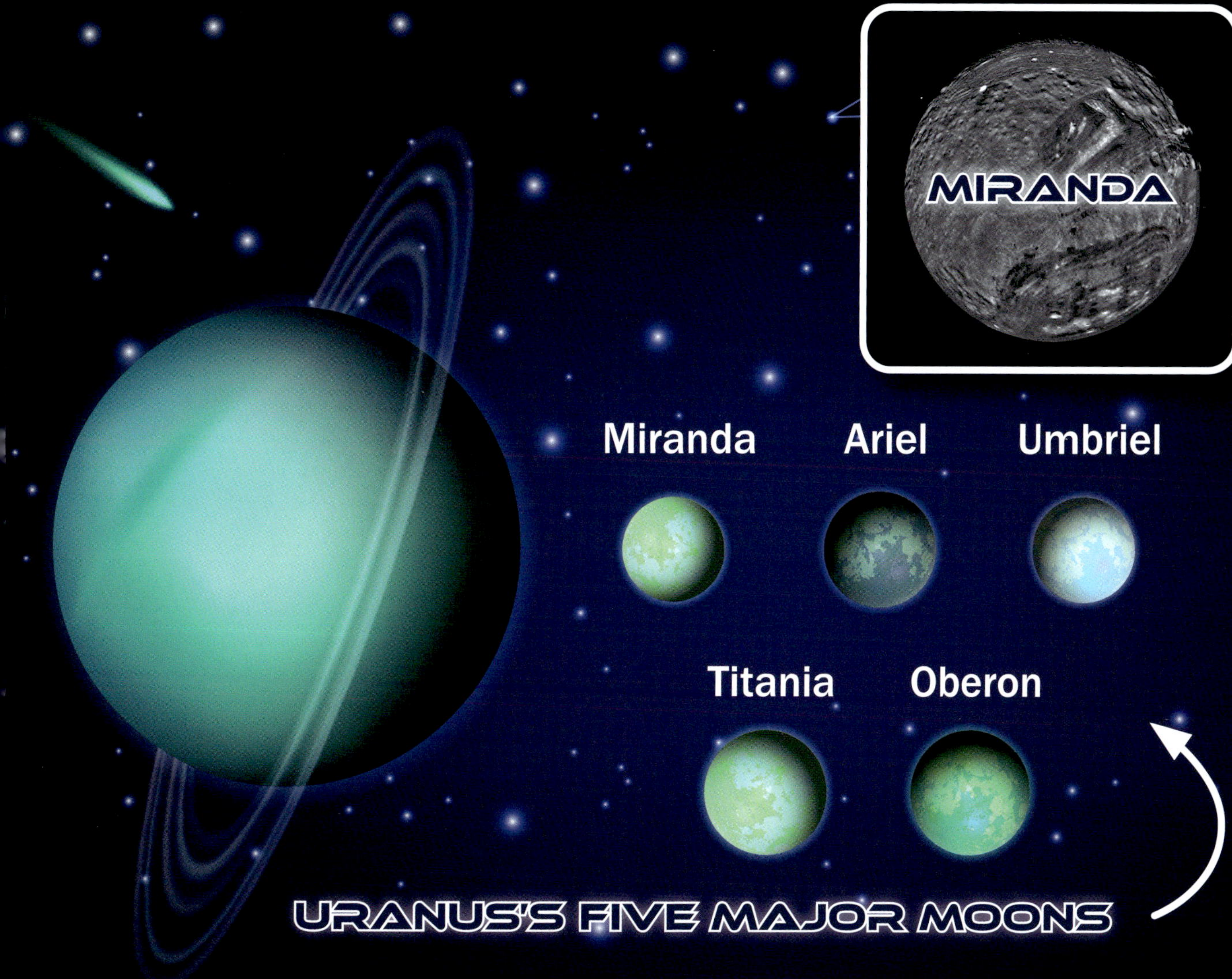

Most of Uranus's moons are named after characters in plays written by William Shakespeare.

Nearly 50 years ago, scientists sent a probe to take photos of Uranus. The probe, called Voyager 2, was the first human-made object to fly past Uranus. Voyager 2 discovered 10 new moons and two rings around Uranus.

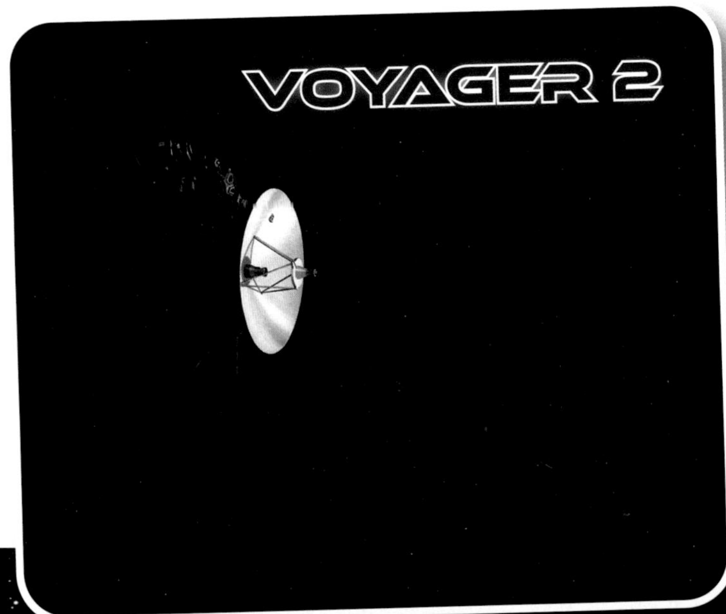

IMAGE OF URANUS TAKEN BY VOYAGER 2

Two twin Voyager probes were sent to take photos of many planets.

WORDS TO KNOW

atmosphere: The mixture of gases that surround a planet.

core: The central part of something.

equator: An imaginary line around a planet that is the same distance from the planet's north and south poles.

methane: A common gas in the solar system. On Earth, it is found in natural gas.

opposite: Going the other way.

orbit: To travel in a circle or oval around something.

probe: An unmanned spaceship.

surface: The top layer of a planet.

surround: To have something in the space all around an object.

telescope: A tool that makes faraway objects look bigger and closer.

FOR MORE INFORMATION

BOOKS

Anderson, Amy and Brian Anderson. *Space Dictionary for Kids: The Everything Guide for Kids Who Love Space*. London: Routledge, 2021

Statum, Hilary. *Solar System for Kids*. Emeryville, CA: Rockridge Press, 2020.

WEBSITES

All About Uranus
spaceplace.nasa.gov/all-about-uranus/en/
Learn about Uranus with the help of photos and fact boxes about the Ice Planet.

Astronomy: Our Place in Space
amnh.org/explore/ology/astronomy
Explore this American Museum of Natural History website with information all about space and the solar system.

Publisher's note to educators and parents: Our editors have carefully reviewed these websites to ensure that they are suitable for students. Many websites change frequently, however, and we cannot guarantee that a site's future contents will continue to meet our high standards of quality and educational value. Be advised that students should be closely supervised whenever they access the internet.

INDEX